How We Know

A Book for the People
Who Want to Know

Jobuydul Islam

To all my favorite faces, who amaze me and bring joy
in every single moment.

Content

Beginning Our Adventure

The Purpose of the Book

"How We Know" is like a treasure hunt through the exciting world of science. It's all about the amazing changes in animals and plants (that's evolution), the incredible story of life on Earth, and how science helps us understand things better than just guessing. This book is a journey for the curious, where you don't need to be a science whiz to join in. Think of it as an adventure into the wonders of our world, just like the famous scientist Carl Sagan once described (Sagan, 1990).

The Importance of Understanding How We Know

Understanding the truth in a world full of confusing stories is super important. Think of science as your superpower to uncover the real story behind what you hear and see. Richard Dawkins, a well-known science writer, said that science reveals the true magic of our world (Dawkins, 2008). And Yuval Noah Harari, another great thinker, reminds us that discovering new things is one of the best adventures (Harari, 2015). Let's dive into the world of science and become detectives in the quest for truth!

Evolution and Natural Selection

What is Evolution?

Evolution is a fascinating journey of change in the world of living things. Imagine a family photo album stretching back over millions of years. In this album, you would see tiny changes in each generation of creatures, gradually leading to the animals and plants we see today. This process is incredibly slow, like a tree growing. It's hard to see the change day by day, but over years, it's clear the tree is taller and stronger. Charles Darwin, a famous scientist, first explained this idea in his book "On the Origin of Species" (Darwin, 1859). He showed that through tiny changes over huge periods of time, all the diverse life we see around us has come to be.

Darwin, Charles. (1859). *On the Origin of Species.*

The Theory of Natural Selection

Natural selection is nature's way of picking the best traits for survival. It's like a nature contest where the prize is getting to live and have babies. This idea was also introduced by Charles Darwin and another scientist, Alfred Russel Wallace. They noticed that in nature, not all animals and plants survive to grow up. Those that do often have something special about them, like being faster, stronger, or better camouflaged. This is not just about animals; plants also compete. For example, some plants might be better at getting sunlight or surviving droughts. The ones that are best at these things are more likely to survive and pass on their traits (Darwin & Wallace, 1858).

Darwin, Charles, Wallace, Alfred Russel. (1858). *Papers presented to the Linnean Society.*

Evidence for Evolution and Natural Selection

There is a mountain of evidence that supports evolution and natural selection. Fossils are one of the key pieces of this puzzle. Fossils are like nature's memory, preserved in rocks, showing us forms of life that existed millions of years ago. Scientists can see how creatures like dinosaurs changed over time into the birds we know today (Lamoureux, 2009). Another big piece of evidence is DNA, which is like a recipe for making a living thing. When scientists compare the DNA of different animals and plants, they find that it tells a story of how all life is connected and has changed over time (Carroll, 2006).

Lamoureux, D. O. (2009). *Evolutionary Creation: A Christian Approach to Evolution.*
Carroll, Sean B. (2006). *The Making of the Fittest: DNA and the Ultimate Forensic Record of Evolution.*

Debunking Creationist Arguments

While some people believe in creationism, which says that all life was created in its current form and doesn't change, science tells us a different story. It's important to respect different beliefs, but when it comes to understanding the natural world, science provides clear evidence of evolution and natural selection. This doesn't mean science conflicts with all religious views. In fact, many religious people also accept the scientific explanation of how life changes (Miller, 1999). It's about understanding that in the realm of natural history, science has gathered a lot of evidence that helps explain the diversity and complexity of life on Earth.

Miller, Kenneth R. *(1999). Finding Darwin's God: A Scientist's Search for Common Ground Between God and Evolution.*

The History of Life on Earth

The Fossil Record

Fossils are like nature's photo album, giving us a peek into the distant past. These are the remains or impressions of ancient organisms preserved in rock. By studying fossils, scientists have put together a timeline of life on Earth. It's like detective work, where each fossil is a clue to what life was like millions of years ago. For instance, dinosaur fossils tell us about these incredible creatures that roamed the Earth long before humans appeared. The fossil record is not just about big animals; it includes tiny plants, insects, and microorganisms too. By studying these fossils, we can see how life has evolved and changed over time (Prothero, 2007).

Prothero, Donald R. (2007). *Evolution: What the Fossils Say and Why It Matters.*

The Age of the Earth

How old is our planet? This is a question that has fascinated scientists for centuries. Today, thanks to modern technology, we know that Earth is about 4.54 billion years old. This age has been calculated using radiometric dating, a method that measures the decay of radioactive elements in rocks. This is like a natural clock hidden in stones. It's not just one rock that tells this story; scientists have tested many rocks from different parts of the world and even from the moon! This helps us understand not just when Earth formed, but also how life on it has had enough time to evolve into the diverse forms we see today (Dalrymple, 2001).

Dalrymple, G. Brent. (2001). *The Age of the Earth.*

The Origin of Species

The origin of species is a fascinating part of Earth's history. It's about how different kinds of animals and plants came into being. Charles Darwin, who we talked about in the evolution chapter, was the first to explain how new species evolve. He proposed that over long periods, small changes add up, leading to the creation of new species. Imagine a family of animals living in a forest. If part of this family moves to a different environment, like a desert, over many generations, they will adapt to this new home. This might lead to them becoming a completely different species from their forest relatives. This process is at the heart of the amazing variety of life we see around us (Darwin, 1859).

Darwin, Charles. (1859). *On the Origin of Species.*

The Role of Religion in Explaining Human History

The Limitations of Religious Explanations

Religion has played a big role in how people have tried to understand the world throughout history. Many religions have stories about how the world was created, how life began, and why things are the way they are. However, these stories are often based on faith and traditions rather than scientific evidence. While these stories are important to many cultures and can teach us valuable lessons about values and history, they don't always match what scientists have discovered about the world. For example, scientific research shows that the Earth is billions of years old and that life evolved over a long time, which is different from some religious creation stories (Scott, 2009). This doesn't mean religion is wrong or bad, but it shows that religion and science have different ways of explaining the world.

Scott, Eugenie C. (2009). Evolution vs. Creationism: An Introduction.

The Scientific Method and Evidence-Based Understanding

The scientific method is a way of finding out about the world that's based on evidence and experiments. Scientists make observations, come up with ideas (hypotheses), test them, and then see if the results support the ideas. This method has helped us learn a lot about the world, from the tiny particles that make up everything to the vast universe around us. Unlike religious explanations, scientific explanations can be tested and proven right or wrong. This doesn't mean science has all the answers, but it does mean that scientific ideas are always being tested and improved (Gauch, 2003). Science helps us build a more accurate picture of the world based on what we can see, measure, and test.

Gauch, Hugh G. (2003). Scientific Method in Practice.

The Separation of Church and State

The idea of keeping religion and government separate is important in many countries. This idea means that the government doesn't support or oppose any religion, and people are free to follow their own beliefs. This separation is important for making sure that scientific understanding and religious beliefs don't get mixed up in laws and education. It helps create a society where people can believe what they want while still making decisions based on facts and evidence when it comes to public policy and education (Jelen, 2007). This separation supports a world where both science and religion can exist without conflicting with each other.

Jelen, Ted G. (2007). *Religion and American Politics: Classic and Contemporary Perspectives.*

The Future of Our Understanding

The Importance of Continuing Research

Continuing research is like keeping the adventure of discovery alive. Every new study, experiment, or exploration adds another piece to the giant puzzle of our world. Just as scientists in the past made discoveries that changed how we understand the world, today's researchers are uncovering new facts and ideas. For example, ongoing research in space exploration is constantly expanding our knowledge of the universe, while studies in biology are revealing more about the complexities of life on Earth. The importance of this continuous quest for knowledge can't be overstated – it's the key to solving mysteries and improving our lives (Kuhn, 1962).

Kuhn, Thomas S. (1962). *The Structure of Scientific Revolutions.*

The Possibility of New Discoveries

Imagine standing at the edge of a vast, unexplored forest full of unknown plants and animals – that's what the future of scientific discovery is like. There's so much we still don't know, and every new discovery can change our understanding of the world. For instance, the discovery of DNA's structure in the 1950s revolutionized biology and medicine (Watson & Crick, 1953). Today, advancements in technology like powerful telescopes and microscopes are helping us see things we've never seen before, from distant galaxies to tiny cells. The possibilities for new discoveries are limitless, and each one has the potential to change the world.

Watson, James D., & Crick, Francis H.C. (1953). "Molecular Structure of Nucleic Acids: A Structure for Deoxyribose Nucleic Acid." Nature.

The Value of Critical Thinking and Skepticism

Critical thinking and skepticism are like superpowers for your brain. They help you question information, look for evidence, and decide what to believe. This is really important in science because it helps make sure that new discoveries are based on solid evidence and not just guesses or wishful thinking. Scientists always question and test their ideas – it's a big part of how science moves forward. By thinking critically and asking questions, we can avoid being fooled by false claims and understand the world more accurately (Popper, 1963).

Popper, Karl. *(1963). Conjectures and Refutations: The Growth of Scientific Knowledge.*

Wrapping Up Our Journey

The Significance of Understanding How We Know

Understanding how we know what we know is incredibly important. It's like having a map in a big city – it helps you find your way and understand where you are. Knowing how science works and how it helps us learn about the world is super useful. It's not just about memorizing facts; it's about understanding how we figure out those facts in the first place. This helps us make good decisions, like choosing the right foods to eat or understanding how to protect our planet. It's also about being smart consumers of information, so we're not easily fooled by things that aren't true (Gilbert, 1991).

Gilbert, Daniel. (1991). *How We Know What Isn't So: The Fallibility of Human Reason in Everyday Life.*

The Responsibility of Educating Ourselves and Others

One of the coolest things about learning is sharing what we know with others. It's our responsibility to keep learning and to help others learn too. When we share knowledge, we help build a smarter, more informed community. It's like lighting candles – each new flame can light many others, and soon, the room is bright. By teaching others what we know about science and how we know it, we encourage everyone to think critically and stay curious. This way, we all grow smarter together and can tackle big challenges that face our world (Feynman, 1985).

Feynman, Richard. (1985). *Surely You're Joking, Mr. Feynman!: Adventures of a Curious Character.*

The Promise of a Brighter Future Through Knowledge

The more we know, the brighter our future can be. Every new thing we learn about science, nature, and the world helps us solve problems and come up with great ideas for the future. Just like planting seeds in a garden, the knowledge we gain today can grow into amazing things tomorrow. Maybe we'll find new ways to cure diseases, protect our environment, or explore distant planets. The possibilities are endless! By staying curious and learning as much as we can, we're not just making ourselves better – we're making the whole world better for everyone (Hawking, 1988).

Hawking, Stephen. *(1988). A Brief History of Time.*

Gratitude to the Great Minds

In our exploration of "How We Know," we've journeyed through the ideas and discoveries of many brilliant minds. It's only fitting to extend our deepest gratitude to these philosophers, scientists, and thinkers whose tireless curiosity and groundbreaking work have illuminated our path.

We thank Charles Darwin and Alfred Russel Wallace for their pioneering work in the field of evolution, which fundamentally altered our understanding of life's diversity. Their courage and insight laid the foundation for generations of scientific inquiry.

Our journey would have been incomplete without the profound contributions of Carl Sagan, whose eloquence and passion for science inspired countless individuals to look at the cosmos with wonder and curiosity.

We are indebted to Richard Dawkins, not only for his evolutionary insights but also for his ability to communicate complex scientific concepts with clarity and enthusiasm.

The profound historical perspectives offered by Yuval Noah Harari have enriched our understanding of humankind's place in the natural world, bridging the gap between history and biology.

Our understanding of the scientific process owes much to Thomas Kuhn, Karl Popper, and other philosophers of science who have delved into the nature of scientific discovery and knowledge.

We also acknowledge the numerous unnamed scientists and researchers whose diligent work continues to expand the frontiers of knowledge. Their discoveries, though they may not always make headlines, are the building blocks of our scientific understanding.

To all these thinkers, past and present, we express our heartfelt gratitude. Their legacies are not just in the pages of history but are alive in the ongoing quest for knowledge and understanding.

Inspirations Behind the Journey

In writing "How We Know," I discovered many amazing books that helped light the way. These books cover everything from science to the story of humans, showing us how we've learned and grown over time. They're like guides on a big adventure of discovery. I've picked out some of these books to share, hoping they'll make you just as excited and curious about the world as they did for me.

"The Selfish Gene" by Richard Dawkins - A groundbreaking exploration of evolutionary biology, introducing the concept of genes as the principal units of selection in evolution.

"The Blind Watchmaker" by Richard Dawkins - Dawkins elucidates the power of natural selection through compelling arguments and thought experiments.

"Sapiens: A Brief History of Humankind" by Yuval Noah Harari - Harari offers a sweeping narrative of human history, from the emergence of Homo sapiens in Africa to the complex societies of today.

"Cosmos" by Carl Sagan - Sagan's masterpiece, blending the science of the universe with the poetry of human exploration and discovery.

"A Brief History of Time" by Stephen Hawking - Hawking's iconic work that brings the complexities of cosmology and theoretical physics to the public.

"The Structure of Scientific Revolutions" by Thomas Kuhn - A profound analysis on the evolution of scientific theories and the paradigm shifts that shape scientific thought.

"Finding Darwin's God" by **Kenneth R. Miller** - Miller's compelling case for the compatibility of evolutionary theory and religious belief.

"Evolution: What the Fossils Say and Why It Matters" by Donald R. Prothero - A robust defense of evolutionary science grounded in the fossil record.

"On the Origin of Species" by Charles Darwin - The seminal work that introduced the theory of natural selection and transformed our understanding of life on Earth.

"Surely You're Joking, Mr. Feynman!" by Richard Feynman - Feynman's entertaining and insightful autobiography, offers a unique perspective on science and learning.

ABOUT THE AUTHOR

Jobuydul Islam, hailing from Rangpur, Bangladesh, transcends his professional identity as a Solution Architect in Oracle Cloud Platform, embracing the essence of a polymath. His academic roots are in Computer Science and Engineering, yet his passions span from the technological realm to mastering the craft of coffee preparation, epitomized by his charming coffee shop, Cafelytics, in Dhaka. An avid reader and traveler, Jobuydul personifies the modern explorer's ethos, perpetually in pursuit of fresh insights and adventures.